A. G Elliott

Industrial Electricity

A. G Elliott

Industrial Electricity

ISBN/EAN: 9783744790499

Printed in Europe, USA, Canada, Australia, Japan

Cover: Foto ©berggeist007 / pixelio.de

More available books at **www.hansebooks.com**

INDUSTRIAL ELECTRICITY

Electro-Mechanical Series

INDUSTRIAL ELECTRICITY

TRANSLATED AND ADAPTED FROM THE FRENCH
OF HENRY DE GRAFFIGNY

AND EDITED BY

A. G. ELLIOTT, B.Sc.

LONDON

WHITTAKER AND CO.

2 WHITE HART STREET, PATERNOSTER SQUARE

NEW YORK: 66 FIFTH AVENUE

1898

EDITOR'S NOTE

THE Editor in presenting this, the first of a series of volumes upon Electro-mechanics, does so with some diffidence; but he believes that there is room for these volumes, because they explain in very clear and non-mathematical language the many and various applications of electricity. In the original French many thousands of these works have been sold, and the Editor trusts that the same appreciation of the volumes may be shown in England and America.

It has of course been found necessary to modify and adapt the original French works to the needs of English readers.

The present volume may be accounted as introductory to the rest of the series. It is divided into short chapters, each dealing with a separate branch of practical electricity. On account of the magnitude of the subject, the information which they supply is necessarily meagre in character, and is only intended to give a general idea of electrical science and its practical applications. The other volumes of the series treat the more important of the branches here touched upon separately and in detail.

April 1898.

CONTENTS

LIST OF ILLUSTRATIONS

INDUSTRIAL ELECTRICITY

CHAPTER I

NATURE OF ELECTRICITY

OUR ideas of the physical world are based on two fundamental conceptions, those of force and matter. Mentally, we are unable to conceive force as apart from matter, and physically it remains an open question whether the two are not in reality but different manifestations of the same entity. We conceive matter as being composed of exceedingly small particles called atoms, arranged in groups called molecules. These molecules are not rigidly connected to one another, and even in the densest matter the intervals which separate them are vast compared with their size, and may be likened to the enormous distances which separate the stars.

Moreover, we have very good reason for believing that there exists, throughout the entire universe, a subtle and all-pervading homogeneous medium, known as the luminiferous ether, which we shall allude to later.

B

Matter can only be perceived through the medium of our senses; thus the more or less rapid movements of the molecules produce the sensations known as heat and cold. A body whose molecules are in more rapid motion than those of our own body appears to us to be hot, while one with slower movements appears to be cold. *Force* may be defined as that which tends to produce motion in matter or to change existing motion.

The atoms constituting matter are not necessarily all of the same size or weight. About seventy different sorts of atoms are known, one corresponding to each of the known elements. These elementary atoms are indestructible, and as far as we know indivisible. Force on the other hand is of one kind only, although manifested to us in such different forms as light, heat, electricity, etc. For instance, those vibrations of the ether known as electricity, remain electricity in passing along a conducting wire, become heat and light in a resisting wire, mechanical work in an electro-motor, and molecular work in an electrolytic cell.

Nothing is lost in nature, nothing is created. The number of atoms and the sum-total of the energy of the universe, are the same to-day as they were when the solar system existed only as an attenuated nebula, and will be the same when the earth shall have accomplished its final destiny.

A vibration in the luminiferous ether may, on reaching a body, either pass through it or be reflected at its surface. In both cases the vibration which leaves the body is not the same as the incident vibration. Thus the reflected ray of light which leaves a mirror is not of exactly the same character as the incident ray. We have a similar phenomenon in the case of heat, sound, and electricity, and this modification of the reflected vibration is called *polarisation*. The phenomena of the polarisation of light have been very carefully studied by means of highly ingenious apparatus devised for the purpose, but less attention has been paid to the polarisation of other sorts of vibration, although they are known to exist.

We therefore see that all the phenomena by which matter is made manifest to us are closely related to motion of some kind or other. Sound, like heat and light, is a vibration, and the researches of Young and Fresnel have shown that space is filled by a medium, capable of vibration, namely the luminiferous ether, of which very little is known, except the mathematical conditions under which waves are propagated in it.

Clark-Maxwell, by the extension of the mathematical theory of light, showed that light and electricity were of exactly the same nature, and Hertz, the German physicist, by his experiments demonstrated the truth of this independently of all mathematical theory.

We can greatly assist our conception of electricity by the use of mechanical analogies. The two-fluid theory which is so helpful in electrostatics as to almost make us forget that it is only an analogy, becomes inconveniently complicated when applied to electro-magnetic problems. Faraday soon rejected the two-fluid theory when he came to consider the action of electric currents, and was the first to suspect the relation of electricity to light. He

himself did not succeed in demonstrating it, but he paved
the way for Maxwell, who proved that light and electric
vibrations are not only propagated by the same medium,
but are also modifications of the same state of the ether.

In order to completely establish the electro-magnetic
nature of light, it was necessary to show that theory agreed
with fact. Hertz succeeded in showing the inverse of the .
above, namely, that electricity was propagated in a vibratory manner, with a definite velocity equal to the velocity
of light.

Maxwell, developing Faraday's ideas concerning the part
played by the di-electric in the propagation of an electric
disturbance, arrived at the conclusion that such a disturbance is propagated with a velocity equal to the ratio of
the electro-magnetic unit of quantity to the electrostatic
unit, and found that this velocity is the velocity of
light.

The effect on an electroscope of discharging a Leyden
jar at a distance of more than 10 metres is inappreciable,
and as the velocity of propagation of the electric disturbance is about 300,000 kilometres a second, it follows that
the 10 metres between the source of the waves and the
electroscope would be covered in about $\dfrac{1}{3 \times 10^7}$ seconds.
It was found impossible to measure, or even to obtain
indications of such a very short interval of time. Cause
and effect seemed to be coincident. Even the time taken
by a Leyden jar to discharge is about 1000 times as long.
Hertz showed that by properly arranging the dimensions
of an electric circuit, a rapidly vibrating discharge could

be obtained whose period was about $\dfrac{1}{100,000,000}$ seconds, and which produced electric waves of less than 10 metres wave-length.

Hertz used for a circuit an induction coil whose secondary terminals were connected to two spheres about 3 c.m. diameter. A difference of potential being set up between the spheres, an oscillatory discharge takes place which gives rise to electric waves. The electroscope consists of a metallic wire, bent into a circle, and terminated at its extremities by a pair of knobs which are brought close together but do not touch. At the instant when a discharge takes place between the large spheres, a very small spark is seen to pass across the gap of the electroscope, due to the rapid variation in the field of the electric force. There is a particular set of dimensions of the electroscope for which the sparks are greatest. This happens when the oscillations which the electroscope might produce itself, have the same period as those of the exciter. It only remains to be stated that an interval of time elapses between the discharge and the induced spark, which shows that the propagation of an electric wave is not instantaneous.

This sort of electroscope which Hertz used is called an electric resonator, for reasons which are obvious. It remains to be shown that electric disturbances give rise to waves which are capable of interference, and therefore that electricity is propagated by periodic vibrations.

The method which Hertz employed is very similar to that used in studying sound-waves. A sound-resonator is moved to different points of a room in which a sound is produced; at certain points in the room the resonator does not act so strongly as in others. This fact alone is sufficient to establish the presence of vibratory motion. .

It is found that two sound-waves of the same length travelling along a line in opposite directions, produce this effect, which is called stationary undulation. The points where the sound is most intense are at equal distances apart, and are called nodes. Savara determined the position of the nodes resulting from the interference of a sound-wave and its reflection travelling back from a wall in the opposite direction. Hertz applied the same method to electric vibrations; his exciter, consisting of an induction coil and spheres, was placed at one end of a large room, with its axis vertical. The opposite wall, about 10 metres off, was covered with zinc-foil connected to the earth. The vibrations produced by the exciter were reflected at this surface and interfered with the waves travelling towards the wall, giving rise to stationary undulations separated by fixed nodes. The wave-length of the vibration is twice the distance between two consecutive nodes. If the time of vibration is also known, the velocity of propagation follows at once. By this method a velocity is found which is very close to that deduced by Maxwell from theoretical considerations, and equal to the velocity of light.

In order to determine whether the waves are longitudinal or transverse, an electric resonator, consisting of a rectangle with two small gaps in its circumference, is placed in the path of waves. If the vibrations are longitudinal, the resonator acts, no matter whether it is placed perpendicular or parallel to the direction of propagation. If, however, the vibrations are executed transversely, the resonator acts in one position only and this last is found to be the case.

If we now place the apparatus for generating the waves at the focus of a concave mirror, a beam of parallel electric rays is projected which can be demonstrated by placing a resonator anywhere in its path. Other properties of light, such as refraction, can be shown to be also possessed by electric waves, prisms of asphalte or pitch being used. Polarisation of electric rays can also be demonstrated by experiment.

We have thus shown that light and electric vibration are one and the same thing, differing only in the lengths of the vibrations. All the proofs of geometrical optics hold equally well for light and electric vibrations, and optics are no longer confined to waves of a fraction of a millimetre in length, but embrace waves of any size whatever, visible or invisible. Light is therefore only electricity which is made manifest to us by one of our senses. Hertz himself wrote : " We henceforth see electricity, where formerly we did not even suspect its presence. The luminosity of every flame or atom becomes an electric phenomenon. Even if a body does not give out light, it may radiate heat and become the centre of electric disturbances. The domain of electricity is extended throughout nature.

Two problems only remain to be solved. Firstly, the mathematical and geometrical conditions of the existence of electro-magnetic phenomena. The solution of this is, at the present date, very nearly complete. Secondly, what is the real nature of the electric and magnetic forces ? It seems as if greater acquaintance with the ether will not only reveal to us its properties, but might also furnish a clue as to the nature of force and mass, which are the very essence of nature. A corner of the veil has been lifted, although many points yet remain to be cleared up, in order to form a fundamental theory of energy. Truth gradually replaces error, and we are gradually beginning to spell the alphabet of nature, and to arrive at a knowledge of the laws which govern the universe.

CHAPTER II

WE have seen that electricity is merely undulatory motion similar to light. It is one form of the energy which is connected with matter, and it is a vibration of one particular order which is produced under certain conditions, and is analogous up to a certain point to the heat of a body.

Before going further it will be necessary to understand the different methods of measuring the motion of electricity. We may consider an electric current as analogous to a stream of water. This analogy is very close, and will give the reader a very good idea how an electric current is formed, and how it flows in a circuit.

In mechanics the potential of a body is a function of the forces which have previously acted on that body. If we consider a mass of liquid under the action of gravity only, the potential of this mass of liquid is proportional to its height above any given level.

If therefore we have two masses of water at different potentials, that is to say at different levels, a current of

9

water will flow from the higher to the lower. If the masses of water are in two tanks, connected by a pipe, the greater the difference in the height of the tanks, the greater will be the pressure of the water at the base of the pipe, no matter what its cross section may be. We may consider the two terminals of a source of electricity as analogous to the two tanks, and the difference of electric potential of the two terminals to the difference of level of the tanks. The current which flows in a wire joining the two terminals is analogous to the current of water in the pipe. Another analogy is that of a hot body connected to a cold body by means of a bar which is a conductor of heat. Heat will flow from the hot body to the cold body, in consequence of the difference in the temperature existing between them. Difference of temperature, level, or potential, are therefore names applied to phenomena which are similar although in reality different.

Since the time of Volta, who was the first to study electric currents, the term *electro-motive force* (E.M.F.) has been applied to that which generates a *difference of potential* between two points of an electric circuit. The electromotive force required to set up a difference of potential between two points is equal to that potential difference. If two chemically different bodies are brought together, an E.M.F. establishes a potential difference between them, which varies in magnitude with different substances. This is the first fact to be noted in voltaic cells.

The current in a circuit always flows from a point at a higher potential to a point at lower potential. If we have a *closed circuit* with a voltaic cell or generator of E.M.F.

included in it, the cell may be looked upon as a sort of pump. The electricity in the outer part of the circuit enters the cell at the negative terminal at low potential, and is pumped up to a higher potential by the cell, before leaving it at the positive terminal. Electricity is to-day an exact science governed by laws founded on a sound mechanical basis. These laws were formulated by a congress of electricians, and are based on the researches of such physicists as Ampère, Faraday, and Ohm. Since 1881 a system of units and measurement has been universally adopted in order to define the various electrical quantities.

This system we will now study.

The *fundamental units* are the *centimetre;* the *gramme-mass*, and the *second*, these being the units of length, mass, and time respectively, from which all other physical and electrical quantities are derived. These three units form what is known as the C. G.S. system (or centimetre, gramme, second), and their quantities are represented by the symbols L, M, and T.

All other units, such as those of surface, velocity, energy, etc., are *derived* from these three, as will be seen from the following definitions of a few of those most used in electrical problems.

The unit of *acceleration* is that of a body whose velocity increases one centimetre per second every second. The acceleration of a body falling under the influence of gravity is about 981 centimetres per second per second. The unit of *force* is that force which gives the unit mass of gramme an acceleration of one centimetre per second per second, and is called the dyne, and is equivalent to the weight of $\frac{1}{981}$ grammes. The unit of *work* is the work done by unit force when its point of application is moved through one centimetre in the direction of the force. It is called the erg. The British engineering units are derived in a similar way from the foot, the pound mass, and the second. The C.G.S. fundamental units are generally too small for practical purpose, and are called absolute units. In practice larger units are used, which have ratios to their respective absolute units which are powers of 10, as the following table shows :

Quantity to be measured.	Name of practical unit.	Symbol.	Number of C.G.S. units in one practical unit.	Multiples.	Sub-multiples.
				Units.	Units.
Electromotive force	Volt	E	10^8	mega = 1 million	deci = 1 tenth
Current	Ampère	I	10^{-1}	myria = 10,000	centi = 1 hund.
Resistance	Ohm	R	10^9	kilo = 1000	milli = 1 thous.
Quantity	Coulomb	Q	10^{-1}	hecto = 100	micro = 1 mil-
Capacity	Farad	C	10^{-9}	deca = 10	lionth
Self-induction	Henry	L	10^2		
Strength of Field	Gauss	G	10^8		

Electro-motive force.—This corresponds to the pressure of water in hydraulics, and is that which tends to produce the current. The *volt* is the unit of E.M.F., and is approximately equal to that of one Daniell cell.

Current.—We measure the current flowing in a wire by the amount of electricity which passes any cross-section in one second, and in a manner similar to that in which we measure the current of water in a pipe by the volume of water which passes any particular cross-section in one second. The *ampère* may be defined as that current which will deposit 1·118 milligrammes of silver per second, or decompose ·09321 milligramme of water in the same time. ·

Resistance is the obstruction offered to the current in passing through a circuit, and is equivalent in the hydraulic analogy to the friction of the water in the pipe. This resistance is measured in *ohms*, and causes a gradual fall of potential along a wire in which a current is flowing, and is the inverse of the conductibility of the wire. The ohm is the resistance of a column of mercury 106 centimetres long and one square millimetre in cross-section. A copper wire 48 metres long and 1 millimetre in diameter has a resistance of about 1 ohm.

The volt, the ampère, and the ohm are connected together by what is known as Ohm's law, namely, that the strength of the current flowing between any two points of a wire is directly proportional to the electromotive force in that wire, or what is the same thing, the difference of potential between the two points is inversely proportional to the resistance between the two points. Kirchoff formulated several laws concerning the flow of currents in closed circuits which are well worth knowing, as they allow the currents flowing in complicated circuits to be easily deduced.

Representing the different units by their symbols given in the table above, we have the following relations:

$$I = \frac{Q}{T} \qquad C = \frac{Q}{E} \qquad I = \frac{E}{R}$$

Quantity.—The unit of quantity is the *coulomb*, and is the amount of electricity which flows in one second past a point in a conductor carrying a current of 1 ampère, and is not to be confounded with the strength of the current.

Capacity.—An electric condenser can be compared to a reservoir for holding gas. The quantity of gas which the reservoir will hold will depend on the pressure of the gas in it and the size of the reservoir. In the same way the quantity of electricity which a condenser will hold depends on its size, and the pressure of the electricity in it. The unity of capacity is the *farad*, and is a condenser which will hold one coulomb of electricity at a pressure of one volt. Practically such a condenser would be of very great size, and much too costly to construct, so commercially a smaller and more convenient unit, the *microfarad*, is used.

Power.—The power, either given out or absorbed by an agent, is the rate at which the agent does work or has work done on it. In the same way, the current in a circuit may be doing work either mechanical, chemical, or otherwise, and the rate at which it does that work is the power absorbed in the circuit. The product of the current and the electro-motive force is the measure of that power. The unit of electric power is called the *watt*, and is that absorbed by a circuit whose resistance is such that an E.M.F. of 1 volt causes a current of 1 ampère to flow round it.

$\therefore$ Power in watts = current $\times$ volts = 10^7 ergs per sec.:

$$\text{or } W = E \times I$$

Watts being measured in ergs per second, the watt-hour, which is the work done in 1 hour by an agent whose power is 1 watt, has been introduced to avoid large numbers. One horse-power is exerted by an agent which lifts 550 lbs. through a height of 1 foot against gravity in one second, and is equal to 746 watts, therefore we have the following relations:

$$1 \text{ watt} = 10^7 \text{ ergs per sec.}$$
$$1 \text{ horse-power} = 746 \text{ watts.}$$
$$1000 \text{ watts} = 1 \text{ kilowatt.}$$

Work or Energy.—The amount of work or energy produced or expended is measured by the number of *joules* or volt-coulombs, and is equal to the watts × time. If no external work is done by a circuit, all the energy of the current is turned into heat, and the quantity of heat evolved in C.G.S. units in a time t is $\dfrac{I^2 R.t,}{J}$ where I is the current flowing, R is the resistance of the circuit, and J is the mechanical equivalent of heat. The C.G.S. unit of heat is the calorie, and is the quantity of heat required to raise 1 gramme of water 1° C. In order to produce 1 calorie of heat we must do work equivalent to raising ·424 kilogramme through a height of 1 metre. This is therefore the mechanical equivalent of heat, and conversely 1 calorie can do ·424 kilogramme-metre of work.

In the next chapter we will tabulate afresh a complete list of all the units used in practical electricity, and we will conclude by giving Kirchoff's laws.

Laws of branched circuits.—(i) In any branching network

of wires, the algebraic sum of the currents in all the wires that meet in any point is zero.

(ii) When there are several electro-motive forces acting at different points of a circuit, the total electro-motive force round the circuit is equal to the sum of the resistances of its separate parts, multiplied each into the strength of the current that flows through it.

We give below an example of a simple circuit. It is necessary in order that a current may flow that the circuit should be continuous, the circuit being then a closed circuit. If there is a discontinuity, no current will flow, and the circuit is then open.

In Fig. 1, N P represents the source of the E.M.F. of the circuit, which may conveniently be taken to be a voltaic element, such as a bichromate cell. We may connect several of these together so as to form a *battery*. One end of the circuit is joined to the positive pole of the battery, and the other end to the negative pole. The circuit is therefore closed through the battery.

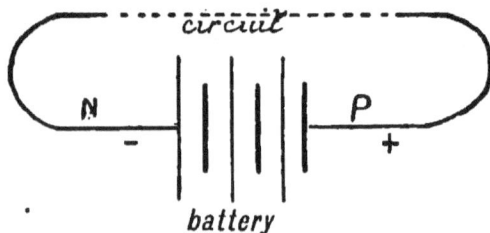

Fig. 1.—Typical Circuit.

The current starts from the positive pole where the potential is highest and flows round the *external circuit* to the negative pole where the potential is lowest, and passes through the battery. While passing through it the current has its potential raised by the E.M.F. of the cells to that of the $+^{ve}$ pole. The path of the current through the cells is called the *internal circuit.*

The external circuit contains all the electrical apparatus on which the current is required to act. The wire coming back from the apparatus is called the *return wire.* In certain cases the earth itself may be used as a return wire. This consideration is of great importance in telegraphy, as it allows of great economy in the length of wire used.

Certain signs and conventions have been adopted in the graphical representation of electric circuits. A cell is represented by a long and thin stroke and then a short thick one drawn at right angles to the conductor, the short stroke signifying the positive pole. The *line wires* are those connecting the source of E.M.F. with the apparatus where the current is made use of.

If we have a number of cells, they can be connected together in two ways. If all the $+^{ve}$ terminals are connected together and all the $-^{ve}$ terminals are connected together the cells are in *parallel.* If the cells are connected in a string, opposite poles being connected together, the cells are said to be in *series.* In a similar manner, any other set of electrical apparatus can be connected up either in series or parallel, or a combination of both.

CHAPTER III

MAGNETISM AND INDUCTION

In certain parts of the world are to be found hard black stones known as lodestones, which possess the curious property of attracting iron or steel, and they are composed of an oxide of iron called magnetite, and their peculiar property is known as magnetism. If a bar of iron, or better still a piece of hard steel, be rubbed with them, it will be found to have acquired the same magnetic property, and it becomes an artificial magnet. If suspended freely at its centre of gravity, such a magnetised bar of iron will always take up a position pointing north and south. The north-seeking end is called the *north* or *north-seeking pole*, and the other end the *south pole* of the magnet. The magnet tends to return to this one position, because the earth itself is a vast magnet whose poles are in the vicinity of the ends of its axis. The line joining the two poles of a magnet is called its magnetic axis.

If we take two magnets and present the north-seeking pole of one to the south-seeking pole of the other, they will be found to attract one another, while if similar poles

18

such as the south poles be presented, they repel one another. The attraction or repulsion is proportional to the product of the intensities of the poles, and inversely proportional to the square of the distance between them. The attraction also depends on the nature of the intervening medium.

If a piece of paper is placed over a magnet and fine iron filings dusted over its surface, the filings settle down into curved lines, showing that the magnetic force is not the same at all points in the space or *field* surrounding the magnet. Any space where there is a magnetic force acting is called a *field of force*, and the lines formed by the iron filings on the sheet of paper are called *lines of force*, and give at each point in the field the direction of the magnetic force at that point. This direction is that in which a small magnetic pole would be urged if placed there.

The lines of force emanating from the north pole of a magnet are considered positive, and those from the south pole negative.

Faraday found that lines of force tend to make themselves as short as possible, and that lines of force of the same sign tend to separate, while those of opposite signs run into one another. The intensity of a magnetic field is measured by the number of lines of force which pass through it, and consequently the force exerted on a bar of iron will be greater according as it cuts a larger number of lines of force. For any given surface the magnetic force on it can therefore be represented by the number of lines which pass through it. If we consider a certain number of lines of force as enveloped by a surface, we obtain a tube of force. If this tube of force have not the same area of cross section at all points, but always contain the same number of lines, the force at any point will be inversely as the area of the cross section of the tube at that point.

Mathematically, the strength of the magnetic field at any point is the resultant of all the forces which would act on a unit pole placed there. The resultant force is equal to the number of lines of force per unit area at the point, and it is in the direction of those lines of force.

This is the same as saying, that the strength of the field depends on the density of the lines of force. The *flux* of force is the total number of lines which pass through any surface, and the strength of magnetisation of a body is the ratio of its magnetic moment to its volume. Bodies which exhibit the phenomenon of magnetisation are called magnetic bodies; such bodies are cobalt manganese of platinum, but the majority of bodies do not show any magnetism of their own, and these are termed paramagnetic bodies. In every case of magnetisation of a body, the lines of force which pass through it, do so by virtue of the magnetising force. This magnetising force may be exerted by the body itself, as is the case in permanent magnets, or it may be due to some external cause, as in electro-magnets, where a current flowing round a coil of wire produces it.

The total number of lines which this force causes to flow through a space is called the flux, and the flux per unit area of the cross-section of the space is called the *induction*, and is denoted by B. The amount of the induction depends upon what material fills the space considered, and this gives rise to a quantity called permeability, denoted by μ, which is the ratio of the induction to the magnetising force producing it. Consequently the greater the permeability of the space, the greater will be the number of lines of force generated in it by the same magnetising force. The permeability of iron gradually increases up to a certain limit of induction, but above that limit it falls rapidly. The permeability of a vacuum is taken as unity, and is practically the same as that of air.

The specific magnetic resistance of a body is the inverse of its permeability, and is called its reluctivity.

Electro-dynamics is that part of the science of electricity which treats of the force which one current exerts on another. Electro-magnetics is that branch which deals with the production of magnetic phenomena by means of electric currents. When a current flows along a wire a magnetic field is always produced in its neighbourhood, the lines of force circulating round and round the wire. We therefore see that the action of currents on one another will depend on the laws which govern lines of force, and that the magnetic field produced by a current is of exactly the same nature as that produced by a permanent magnet.

Ampère, the celebrated French physicist, discovered the following laws :—(1) Two parallel wires having currents flowing in the same direction attract one other, and if the currents are in opposite directions they are mutually repelled. (2) Two portions of circuits crossing one another at an angle, attract one another, and tend to become parallel if the currents both flow towards or away from the apex of the angle. They repel one another if one current approaches the apex while the other recedes from it. (3) The force exerted between two parallel portions of circuits is proportional to the products of the strengths of the two currents and the lengths of the portions, and so inversely proportional to the distance between them. The whole theory of dynamo-electric machinery is founded on these laws. The following laws are also important :—

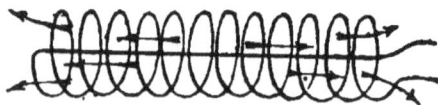

FIG. 2.—Solenoid.   FIG. 3.—Electro-magnet.

Maxwell's Law.—If a movable circuit with a current flowing in it is brought into the field of a magnet, it tends to take up a position, embracing the maximum number of lines of force possible. By means of Laplace's law we are also able to determine the strength of field produced by an element of circuit at a point in its neighbourhood.

Electro-magnets.—A spiral coil of wire which has a current flowing in it, and is without any iron core, is called a solenoid (Fig. 2). In consequence of the current, a magnetic field is produced within the spiral, whose lines of force are parallel to the axis of the coil. Such a solenoid behaves exactly like a magnet, so that if we present the pole of a powerful permanent magnet to one end of the coil, it is repelled or attracted, depending on the similarity or dissimilarity of the poles.

The only difference is that in a solenoid the poles are at the ends, while in a bar magnet they are at a short distance from them.

If we wind several layers of insulated wire on a bobbin with an iron core we have an electro-magnet. The lines of force generated by the solenoid pass through the iron, which has a very much greater permeability than air, consequently a greater number of lines of force are induced in the iron than would have been if there had been no iron core.

In practice, an electro-magnet consists of a rod of soft annealed iron, bent into the shape of an U, or else two rods riveted to a flat iron base or yoke : the two rods form the core of two bobbins which are slipped over them. The bobbins are wound with silk or cotton-covered copper wire, which is more or less thick according to the use to which it is to be put (Fig. 3).

Lenz and Jacobi gave the following laws, which are only true for small electro-magnets :—

The strength of an electro-magnet is proportional to (1) the current; (2) the number of turns of wire in the coil; (3) the square root of the diameter of the core.

If we consider a continuous line drawn such that at every point its tangent is parallel to the direction of the magnetic force, the work done in moving a unit once round this line is the *line integral* of magnetic force round it, and is called the *magneto-motive force* in this *magnetic circuit.*

The induction in a bar of iron depends on the strength of the field in which it is placed, but this law is not an exact one, because a bar subjected to a variable magnetising force, exhibits a lag of magnetisation when it is being demagnetised. This lag is called hysteresis, and its effect is to heat the bar of iron with consequent loss of energy. This loss of energy has to be taken into account in practice, especially in magnetic apparatus and dynamos.

Induction.—If we move a conducting wire in a magnetic field an E.M.F. is generated in it, causing a current to flow called an induced current, which lasts only as long as the movement continues. Magnets and currents both give rise to magnetic fields, which in their turn may generate induced currents. These phenomena were studied by Faraday, and called the phenomena of induction.

If we have a circuit acting inductively on a second circuit, the two are called *primary* and *secondary* circuits respectively, and the currents in them are called primary and secondary currents. An induced current is produced whenever the flux of lines passing through a circuit is either increased or decreased. An induced current may be therefore produced in three separate ways: by the action of currents, by the action of magnets, and by the action of the earth's magnetic field.

It is not, strictly speaking, a current which is induced in the circuit at all, but an E.M.F. which is generated by the changing of the flux passing through the circuit, which may or may not produce a current according as the circuit is *closed* or *open*. Self-induction is the inductive effect of a circuit on itself.

Maxwell enunciated the following principle: if a closed circuit is displaced in any magnetic field in such a way as to alter the number of lines of force passing through the circuit, a current is generated in it which lasts only as long as the flux is changing.

If we suppose the positive direction of lines of force to be that along which a free N pole would move, and the positive direction in a circuit to be the same as the direction in which the hands of a clock move, then we may say that if the flux enters the circuit in a $+^{ve}$ direction, any diminution of that flux will produce a current in the positive direction round the circuit.

The following law was discovered by Faraday: if a straight conductor moves in a magnetic field so as to cut lines of force at a given rate, a difference of potential is generated between its ends, which is proportional to this rate, if the conductor is and moves normally to the direction of the lines of force.

Lenz found that if we have a relative displacement between a conductor and a magnetic field, the induced current is such as to tend to prevent the motion. If we combine this last with Maxwell's law, we see that the direction of the induced current for any variation of flux is such as to oppose this variation by its own action on the flux. If we divide the time during which the induced current is flowing into a number of small intervals, then the quantity of electricity which passes is equal to the sum of the small intervals, each multiplied by the current which flowed during that interval.

If two circuits are close to one another a portion of the flux generated by one will pass through the other. The *co-efficient of mutual induction* of the two circuits is the number of lines due to the first circuit, which also pass through the second circuit when unit current flows through the first circuit. The phenomenon of *self-induction* is absolutely analogous. If we suppose the two circuits above to approach one another so as to actually coincide, the mutual induction becomes a self-induction of the circuit. The mere fact, therefore, of a current in a circuit generating lines of force, creates an E.M.F. in that circuit, which is in a contrary direction to the E.M.F. which produced the current. This back E.M.F. of self-induction is proportional to the number of lines of force which unit current would produce when flowing round the circuit. This definition is only true when the permeability of the medium surrounding the circuit is constant. We will now give a table recapitulating all the terms so far established.

ÉSUMÉ OF UNITS AND DIMENSIONS OF ELECTRIC QUANTITIES

Physical quantity.	Symbol.	Dimensions.	Name of C.G.S. unit.	Practical unit.	English unit.
Fundamental.					
Length	L	L	centimetre	metre	foot
Mass	M	M	mass of the gramme	mass of kilogramme	mass of pound
Time	T	T	second	second, hour	second
Geometrical.					
Surface	S or A	L^2	square cm.	square metre	square foot
Volume	V	L^3	cubic cm.	cubic metre, litre	cubic foot
Angle	a, β, θ	—	radian	degree, minute, and second	degree, minute, and second
Mechanical.					
Velocity	v	$L\,T$	cm. per sec.	metre per sec.	foot per sec.
Angular velocity	ω	T	radian per sec.	revolution per min.	revolution per min.
Acceleration	a	$L\,T^2$	cm. per sec., per sec.	metre per sec., per sec.	feet per sec., per sec.
Force	F	$L\,M\,T^2$	dyne	gramme kilogramme	poundal
Energy, work	W	$L^2\,M\,T^2$	erg	kilogramme metre	foot pound
Power	P	$L^2\,M\,T^3$	erg per sec.	kilogramme metre per sec.	foot pound per sec., or H.-P.
Pressure	p	$L\,M\,T^2$	dyne per cm.^{2}	kilogramme per square centimetre	pound per square inch
Moment of Inertia	I	$L^2\,M$			
Electro-magnetic.					
Resistance	R, r	$L\,T$	—	ohm	ohm
Electro-motive force	E, e	$L^{\frac{3}{2}}\,T^{-2}\,M^{-\frac{1}{2}}$	—	volt	volt
Potential difference	V, v	$L^{\frac{3}{2}}\,T^{-2}\,M^{+\frac{1}{2}}$	—	volt	volt
Current	i	$L^{\frac{1}{2}}\,M^{\frac{1}{2}}\,T^{-1}$	—	ampère	ampère
Quantity	Q	$L^{\frac{1}{2}}\,M^{\frac{1}{2}}$	—	coulomb	coulomb
Capacity	C	$L^{-1}\,T^2$	—	farad and microfarad	farad and microfarad
Electric energy	W	$L^2\,M\,T^{-2}$	—	joule	joule
Electric power	P	$L^2\,M\,T^{-3}$	—	watt, kilowatt	watt and kilowatt
Specific conductivity	δ	$L^{-2}\,T$	—	ohm per centimetre cube	ohm per centimetre cube
Self-induction	L	L	—	Henry	Henry
Magnetising force	H	$L^{\frac{1}{2}}\,M^{\frac{1}{2}}\,T^{-1}$	—	ampère turn per centimetre	ampère turn per centimetre
Magneto-motive force	F	$L^{\frac{1}{2}}\,M^{\frac{1}{2}}\,T^{-1}$	—	"	"
Magnetic.					
Intensity of pole	m	$L^{\frac{3}{2}}\,M^{\frac{1}{2}}\,T^{-1}$			
Intensity of field	B	$L^{\frac{1}{2}}\,M^{\frac{1}{2}}\,T^{-1}$			
Flux of force	Φ	$L^{\frac{3}{2}}\,M^{\frac{1}{2}}\,T^{-1}$		No special unit.	
Magnetic induction	B	$L^{\frac{3}{2}}\,M^{\frac{1}{2}}\,T^{-1}$			
Permeability	μ	—			
Magnetic reluctance	L	L^{-1}			

CHAPTER IV

PRACTICAL MEASUREMENT OF ELECTRICAL QUANTITIES

THE methods of measuring electrical quantities may be divided into two classes, according as they are compared with similar quantities, or have their value deduced from their effect on various other known quantities. To the first class belong all methods of opposition, substitution, and comparison : all measuring instruments are based on one or other of these two principles. Instruments for measuring currents are of several kinds, the most simple being the *galvanometer*, whose action is electro-magnetic, consisting of a magnetised needle freely suspended inside a coil. The needle takes up a position which varies with the strength of the current flowing round the coil.

We will give a few examples illustrating those most generally met with in practice.

Tangent galvanometer.—In this instrument the current in the coil is proportional to the tangent of the angle of deflection. The scale is of circular shape, and must always be placed in the same position relative to the magnetic meridian. The needle consists of a very small magnetised bar, suspended at the centre of a large circular coil, with a fine pointer attached to it in order to magnify the deflection produced, and thus increase the sensibility of the instrument.

Sine galvanometer.—In this type the deflecting force is proportional to the sine of the angle of deflection, and the arrangement of needle and scale is the same as in the preceding case.

Astatic galvanometer.—This instrument was invented by Nobili. The needle is made up of two, suspended parallel to one another so as to increase the sensibility by diminishing the controlling force of the earth's field. The needles carry a light pointer which moves over a finely divided scale, allowing the deflections to be easily read.

Reflecting mirror galvanometer.—The reading of very small deflections of the needle was always a source of great trouble until Lord Kelvin invented his reflecting galvanometer, which allows exceedingly minute deviations of the needle to be easily detected and measured. To the magnetised needle is attached a light mirror on which falls a beam of concentrated light ; this beam is reflected and forms a spot of light on a scale some distance off. A very small motion · of the needle may be made to produce a large deflection of the spot. Some of these galvanometers are constructed so as to be dead-beat, that is to say, a small aluminium vane is attached to the needle, which offers a considerable resistance in air friction to sudden movements. The spot of light therefore quickly takes up its final position, without first oscillating to and fro.

D'Arsonval dead-beat galvanometer (Fig. 4).—This galvanometer is essentially different from those which have preceded it, in that the controlling magnet is fixed, whilst the coil through which the current flows is movable. This coil is stretched tight between the poles of a powerful permanent magnet, by two fine metallic threads, one above and one below. These metallic suspensions also serve the purpose of conveying the current to the movable coil. These galvanometers are usually made reflecting.

Fio. 4.—D'Arsonval Galvanometer.

All the galvanometers hitherto discussed are constructed to carry none but very small currents. If we require to measure a large current which might damage these instruments, a known fraction of the whole current is allowed to pass through it. The galvanometer is then said to be *shunted*. This shunting of the instrument is usually effected by means of coils of known resistance enclosed in boxes, which only allow $\frac{1}{10}$, $\frac{1}{100}$, or $\frac{1}{1000}$th part of the whole current to pass through it. The whole current is therefore measured, although only a portion of it passes through the galvanometer.

Electro-dynamometers.—The principle of these instruments is based on the action of one current on another. The current to be measured passes through two coils, one of which is fixed, and the other, a movable coil, is suspended by a torsion head. When a current passes through, the force with which the coils attract one another is measured by the amount of torsion produced, which is proportional to the square of the current. In the best Siemens' dynamometers, currents varying from 1 to ·001 ampère can be measured.

Galvanometers are not suitable for the measurement of currents in commercial applications of electricity, and they have been replaced by another type of instrument which has a needle moving over a scale, from which volts or ampères can be read off directly. These *voltmeters* and *ammeters* stand in the same relation to the generator of electricity as the pressure-gauge does to the steam-engine. They are therefore indispensable where electricity is generated on a large scale, and much time and ingenuity has been expended on the construction of the reliable and accurate instrument, now to be found in every well-equipped electrical installation. We will now proceed to describe some of the best-known types for use in both direct and alternate current work.

Davies' unipolar voltmeter (**Fig. 5**).—This instrument consists principally of a rectangular coil of wire moving round a single magnetic pole and rotating around one of its sides as axis. A permanent magnet maintains a uniform field of force in a narrow annular gap. The magnetic circuit being a nearly closed one (the thickness of the air gap is less than $\frac{1}{8}$ in. and the cross sectional area of it about 3 sq. in.), the reluctance of the circuit is

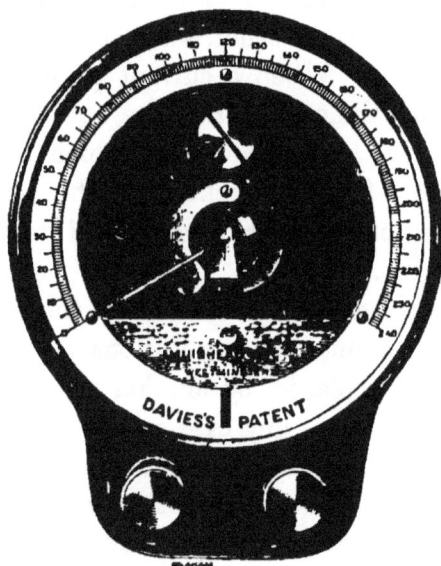

FIG. 5.—Davies' Ammeter (Muirhead and Co.).

low, so that permanency of magnetic induction is well assured, and an instrument is produced giving uniform scale divisions throughout. All forms of this instrument having spiral springs for the directive force have a wide range of action, the scale divisions forming an arc of about 240°. It is made in a vertical form for wall purposes, and in a horizontal form for bench use, reading accurately from 0·01 volt or lower, up to 2·5 volts; scale marked in ·01 divisions, on a 6-inch dial.

Messrs. Muirhead and Co. also construct ammeters on the same principle.

Nalder Brothers' ammeters and voltmeter (Fig. 6).—The base of these instruments is of cast-brass, turned, polished, and lacquered. On this are cast bosses to receive the working parts. The coil is mounted on ebonite, so as to give high insulation, and is screwed direct to the base.

Fig. 6.—Ammeter by Nalder Brothers and Thompson.

It is wound with a thick wire in the ammeter and a fine wire in the voltmeter. For the ammeter the ends of the winding are brought to heavy brass lugs, which pass through the base of the instrument, and are arranged for socket connections. These lugs are carried on ebonite, thus ensuring high insulation. The windings for ammeters, up to 500 or 600 ampères, is of braided cable, and the ends are sweated very carefully into the lugs to ensure good contact. The dial is carried on the bobbin itself, and is therefore highly insulated from the base.

D

The cover fits to the base, but does not touch the dial as shown in the figure by the black line round the bevel. This ensures perfect insulation of the case from the winding, and it is quite safe to lay one's hands on the voltmeters when measuring 2000 volts. The needle is a specially shaped piece of soft iron fixed on a light shaft with steel pivots at either end. The pivots turn in two polished sapphire centres. The ammeters have a uniform scale the whole way up, but in the voltmeters the scale widens out at the working pressure for which the instrument is designed.

Some voltmeters and ammeters are made so as to record their indications on a paper band, which is drawn by clock-work past a style attached to the end of the needle. The paper band has lines printed on it forming a scale of volts or ampères, ruled so as to be concentric with the axis of the needle. The band is usually placed round a cylinder, which is rotated by clock-work or by a mechanism regulated by the current itself. The cylinder may be made to rotate once in any convenient period. An example of such a recording voltmeter is given in Fig. 7.

This instrument is constructed so as to measure alternating as well as direct differences of potential. Its action is not electro-magnetic, but is founded on the fact that metals expand when heated. The terminals of the voltmeter are connected to a long and very fine wire, which becomes heated by the current flowing through it. This wire is passed over a series of pulleys so as to take up as little space as possible; one end of the wire is fixed, and the other is attached to the needle carrying a style at its end. When a difference of potential is applied to the terminals, the wire becomes longer in consequence of the heating due to the current flowing in it, and this

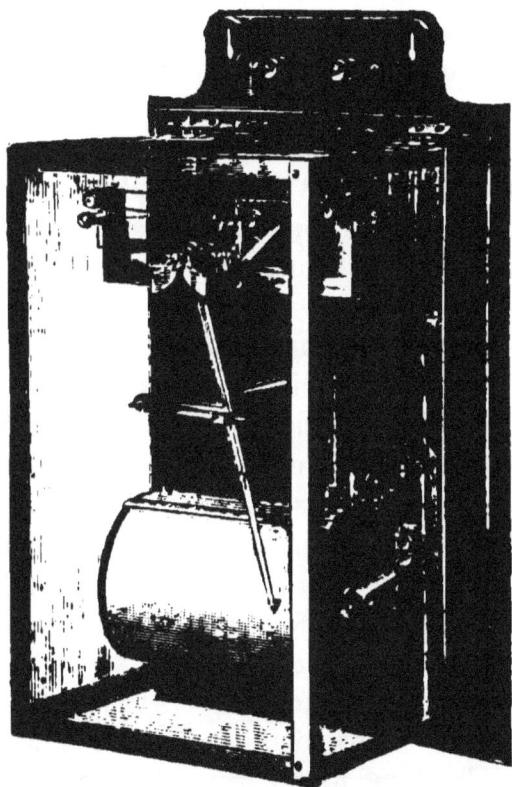

Fig. 7.—Holden Hot-wire Voltmeter.

allows the needle and style to move across the paper scale. Recording ammeters have been constructed on the same principle, by attaching a style to the end of the needle. It is very necessary that all these instruments should be dead-beat, that is to say, the needles must come quickly to rest, and not vibrate to and fro about their mean position. This is attained by various means; sometimes by a light vane, which, placed at right angles to the plane of motion of the needle, tends to oppose any rapid movement through the air; also by attaching an aluminium vane to the needle in its plane of movement, and placing a small, powerful permanent magnet attached to the frame of the instrument, so that the vane moves in a powerful magnetic field, which opposes any rapid motion, by the production of Foucault currents in the metal of the vane itself.

Ayrton-Mather electrostatic voltmeter (R. W. Paul, maker). —This type of voltmeter is rapidly gaining favour among electrical engineers, especially for alternating currents. It possesses the advantage over the ordinary hot-wire voltmeters, that it does not use up any power. The action of the instrument depends on the electrostatic attraction or repulsion between two conductors connected with the two points whose potential difference it is required to measure. In some instruments the conductors take the form of flat plates, but in the Ayrton-Mather voltmeter they are shaped into parts of cylinders as shown diagrammatically in Fig. 8. The three fixed curved plates A, B, and C, are firmly connected together by the plate D, while the needle, consisting of the curved aluminium plates E, F, fixed to the staff G, is drawn into the space between the plates A, B, and C, moving a long pointer over its scale. The controlling force is gravity acting on a weight attached to the needle. The glass in front of the dial is coated on the inside with a transparent conducting varnish, so as to entirely shield the instrument from external electrostatic action. By properly altering the width of the plates the scale of the voltmeter can be opened out to any required degree at its centre, the normal working pressure being an approximately fixed point.

It is necessary in all galvanometers, whether ammeters or voltmeters, that the working parts should be as light as possible, so as to reduce their moment of inertia and make them move quickly. In a central station the instruments, such as switches, voltmeters, ammeters, etc., are all

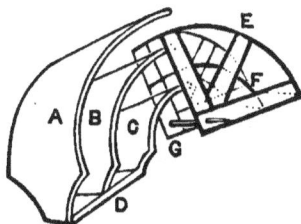

Fig. 8.—Ayrton-Mather Electrostatic Voltmeter. Diagrammatic sketch of the Needle and Inductors.

arranged on a large switch-board, so that the attendant in charge can see at a glance whether the pressure of the mains is correct, and that none of the dynamos are overloaded. Several ingenious arrangements have been invented for automatically giving an alarm if the pressure is too high or too low, or if the current in any circuit is too great, by the ringing of bells, lighting of coloured lamps, etc.

By means of these various instruments the engineer can instantly tell exactly what the plant is doing, and always has complete control over the electrical output of the station. By the use of recording ammeters, a curve is obtained on the paper, which, supposing the potential difference of the mains to be constant, shows the exact load on the station for every moment of the day, the knowledge of which is a very important factor in the economy of an electric lighting station.

Wattmeters.—Recording wattmeters are sometimes more advantageously used than ammeters if the current exceeds 2000 ampères. They measure the product of the volts and ampères, and usually consist of one long thin and one short thick coil without any iron cores, in the same way as dynamometers.

Measurement of resistances.—The operation which has most frequently to be performed in all classes of electrical work is the measurement of resistance. The necessary apparatus for such measurements consists of a battery of cells, a galvanometer, and a number of coils whose resistances are known. In order to facilitate the measurements, the coils are, as a rule, grouped together in boxes, and they can be intercalated in the circuit by simply inserting metallic plugs. These standard resistances are usually made of German silver wire, or an alloy of platinum and silver, covered with silk and wound on bobbins, the whole being saturated with paraffin wax to ensure good insulation. The two ends of one of the coils are connected respectively to two brass blocks, and each block can be electrically connected to its neighbours by inserting a brass plug between them.

The coils are so arranged that resistances of 1, 2, 3, 4, 5, 10, 20, 50, 100, 500, 1000, 2000 ohms can be introduced into the circuit, by combining the given resistances in different ways.

The following method, devised by Christie and applied by Wheatstone, to measure resistances is most frequently used, and is known as Wheatstone's Bridge. The circuit of a battery is made to divide at A (Fig. 9) into two branches, A B C and A D C, which re-unite at C. The current in the circuit is therefore divided in two portions, which flow along A B C and A D C, called the arms of the

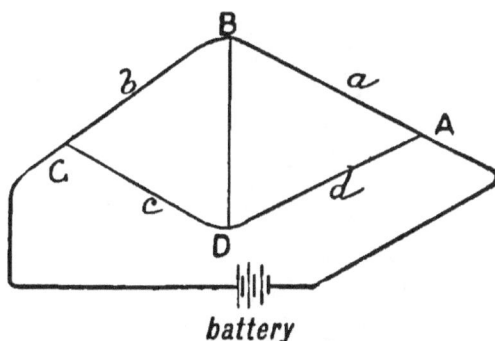

battery

Fig. 9.—Wheatstone's Bridge

bridge. If we consider two points B and D in the two circuits, and join them by a galvanometer, they will not, in general, be at the same potential, and consequently a current will flow through the galvanometer. It is possible, however, by moving one of the points, say B, along the conductor to C, to find a point in it which is at the same potential as D. When this is the case, no current will flow through the galvanometer, and it may be shown that the ratio of the resistances A B and B C is equal to the ratio of the resistances A D and D C. If therefore A D is an unknown resistance, we can determine it by giving the proper values to the resistances C D, A B, and B C.

It is usual in practice to assemble for convenience all the resistances necessary for making a measurement into one box. Various arrangements of the coils in such resistance boxes are in use: the most general is to have resistances of 1, 2, 2, 5, 10, 20, 20, 50, 100, 200, 200, 500, from which any whole number of ohms from 1 up to 1000

FIG. 10.—Resistance Coils arranged for use as a Wheatstone's Bridge.

can be obtained. These correspond to the resistance C D in Fig. 9. In addition to these resistances, another set are provided in the box, for obtaining a suitable ratio for the arms, corresponding to A B and B C in Fig. 9. They consist of two sets of 10, 100, and 1000 ohms, from which ratios of $\frac{1}{1}$, $\frac{10}{1}$, and $\frac{100}{1}$ can be obtained. In another arrangement (Fig. 12), these last two sets of resistances are replaced by a long wire T T' provided with a sliding piece d, which corresponds to the point B in Fig. 9.

Standards of electro-motive force and resistance.—We have previously stated, that differences of potential are measured by means of voltmeters. These voltmeters

Fig. 11.—Resistance Coils.

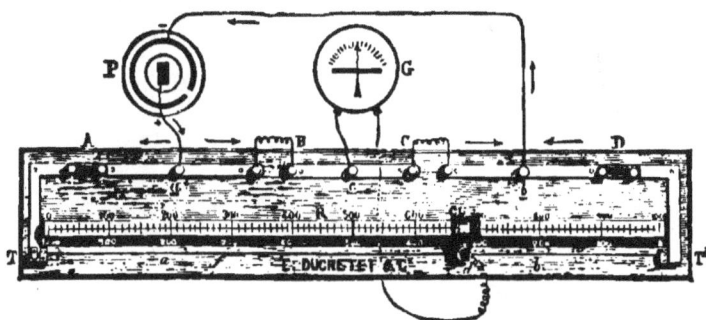

Fig. 12.—Slide Wire form of Bridge.

must be calibrated by means of standards of electro-motive force. No exact unit standard of electro-motive force has yet been constructed, and for this purpose electricians have been obliged to use primary cells whose E.M.F. is very

accurately known, such as Latimer Clarke's, the Post-
Office, Warren de la Rue's, and other cells. Latimer
Clarke's cell (Fig. 13) was adopted by the Chicago Congress,
as affording a fairly accurate standard volt, and consists of
two glass tubes, with a connection between them. In one

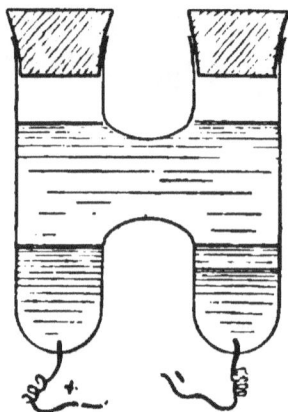

Fig. 13.   Standard Clarke Cells.   Fig. 14.

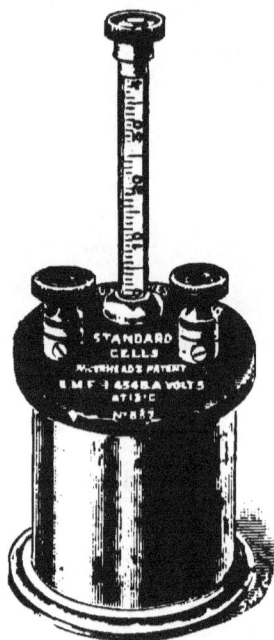

of the tubes is placed an amalgam of pure mercury and
zinc, and in the other, pure mercury with a layer of
sulphate of mercury on top of it. The cell is then filled
up with a saturated solution of sulphate of zinc, into
which a crystal or two of the zinc sulphate is dropped to
prevent super-saturation. The tubes are sealed up air-

tight with plugs of paraffin wax, and the poles are formed by platinum wires, which are fused through the bottom of each tube, and make contact with the mercury. The E.M.F. of such a cell is about 1·35 volts at 15 degrees Centigrade.

The Post Office standard is simply a Daniell cell in an ebonite box. A zinc strip is immersed in a solution of zinc sulphate, surrounding a porous pot containing a copper plate immersed in a strong solution of copper sulphate. At the bottom of the porous pot is placed a small stick of zinc, on which is precipitated any copper from the sulphate which might have passed through the porous pot; in this manner the two liquids do not mix, and the cell gives a very constant E.M.F. of about 1·08 volts on an open circuit.

The Warren de la Rue standard cell consists of a glass tube containing a solution of chloride of ammonia, in which is immersed a stick of zinc and a porous pot made of parchmentised paper. This pot contains a silver wire with silver chloride fused round it, and is closed by a plug of paraffin wax. This cell gives an E.M.F. of about 1·05 volts.

The unit of resistance is the ohm. A standard resistance usually consists of a German silver wire, wound on a bobbin, and enclosed in a metallic box. The ends of the wire are connected to massive copper blocks, which are sometimes formed into cups to hold mercury. Into these mercury cups are plunged the ends of the wires going to the rest of the apparatus in use. The resistance of such a standard is not constant, but varies with its temperature. In order to make accurate use of such a standard, it must

therefore either be kept at the temperature at which the
maker guarantees it to be correct, and which is marked
on the box, or a correction must be made for the temperature at the moment of the experiment.

Measurement of electric energy and power.—We must
remember that electric power measured in watts is the
product of current in ampères, and the difference of potential measured in volts.   In practice this product is obtained
by means of a special apparatus called a wattmeter, in
which there are two coils, one consisting of a short thick

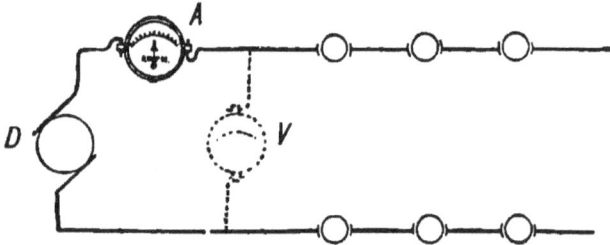

Fig. 15.—Method of joining up a Voltmeter and an Ammeter in a circuit.
V, Voltmeter. A, Ammeter. D, Dynamo.

wire and the other of a long thin wire.   The current to be
measured passes through the thick wire, while the long
wire is connected to the two points, in between which it is
required to measure the power developed.   The deflection
of the needle of these instruments is proportional to the
product of the currents in the two circuits, but as the long
wire has a large resistance, the current flowing through it
is proportional to the difference of potential between its
ends.   The deflections therefore indicate the product of
the current and the electro-motive force, that is to say, the
power,

Electric energy is power multiplied by time. It can be measured by means of a recording wattmeter, based on the principle we have just explained, and in which the needle carries a style. A curve can thus be traced on a paper cylinder moved by clock-work, the area of the curve traced measuring the total energy expended or absorbed.

A great number of the electric energy meters supplied by the electric light companies to their consumers, are simply wattmeters, which record their own indications at short intervals by means of clock-work, as in the meters of Cauderay and Déjardin, but other and more accurate systems have been devised, such as the Thomson-Houston, Bellić, and other meters, in which the rapidity of motion of the recording apparatus is regulated by the current. In the Aron meter there are two pendulums regulated so as to have the same period of oscillation. On of them is an ordinary pendulum, and the other is formed by a steel magnet, and above this magnet is a solenoid through which passes the current to be measured. Each pendulum is moved by a spring and clock-work, and a differential mechanism which they both control moves the recording needle.

Measurement of quantity of electricity.—Quantity of electricity is measured by the intensity of a current, multiplied by the time during which that current flows. The measurement of quantity is usually effected by means of an apparatus known as a *Voltameter*, in which the decomposition of conducting liquid is brought about by the current. Various systems have been devised for this purpose, notably by Edison, Lippmann, Busch, and others.

The Edison voltameters are used as a shunt, and only take a known fraction of the whole current to be measured. The conducting liquid was copper sulphate in the early ones, but this has been replaced by zinc sulphate in the latest types, which are used as energy meters. The plates which the apparatus contains are weighed once a month, and the number of coulombs of electricity which have passed, is calculated from the gain in weight, and from the fact that one ampère hour (3600 coulombs) deposits 1·228 milligrammes of zinc. In other types by the same inventor, the plates become reversed by an automatic movement when their weight reaches a certain value; the plate which formerly was anode becomes cathode until they are again reversed. The number and period of the reversals is measured by a counter, and a simple calculation allows the quantity of electricity which has passed to be measured. These meters are much used in America.

There still exist voltameters with silver electrodes immersed in a solution of silver nitrate. The metal of the positive plate is dissolved by the action of the current, and an equivalent quantity is deposited on the negative plate. The plate is weighed before and after the experiment, and the quantity of electricity which has passed through the apparatus is calculated on the basis that one coulomb (one ampère second) deposits ·00118 gramme of silver.

The above constitute the principal apparatus and methods used in the electric industry, for determining the value of the different quantities which make up any electric current.

CHAPTER V

Theory.—If we take a vessel containing brine or weak sulphuric acid, and place in it a plate of copper and a plate of pure zinc, as long as the two plates do not touch one another nothing happens, but if we join them by a wire the zinc will be immediately attacked by the acid. Bubbles of hydrogen are also liberated on the copper, and if we bring a compass needle near the wire it is deflected, showing that a current is flowing in it. This phenomenon may be explained by considering the molecules of the sulphuric acid in solution, and imagining the equilibrium of the sulphuric acid molecule to be destroyed by its solution in water. The molecule is thus split up into two, one containing hydrogen and the other sulphur and oxygen: these are called ions. When a current flows the sulphur and oxygen attack the zinc and form zinc sulphate, while the hydrogen ions are deposited on the copper plate, and the energy liberated by the action on the zinc supplies the energy of the current. If we now break the connection between the two plates, the electric

47

action immediately ceases, the equilibrium of the molecules being restored. The plates themselves, however, remain at different potentials, so that on joining them by a conductor, a current flows from one to the other, and the strength of current which flows will naturally depend on the chemical affinities of the substances concerned, the area of the surface of metal attacked by the acid, and the total resistance of plates, liquid, and conductor.

A current will continue to flow until either all the zinc or sulphuric acid has been used, that is, assuming that the hydrogen generated at the copper plate is continuously removed; gases being very bad conductors of electricity, a film of hydrogen bubbles offers increased resistance to the flow of current. A film of hydrogen also diminishes the effective E.M.F. of the cell by setting up an opposing electro-motive force, for hydrogen is almost as easily attacked as zinc, especially when freshly deposited. If the film of hydrogen is therefore allowed to remain on the surface of the plate, the cell is said to become *polarised*. We will explain later the various means adopted in order to neutralise the effects of polarisation.

History.—In the year 1799 Volta, professor in the University of Pavia, devised the first chemical generator of electricity. It consisted of columns of alternate zinc and copper discs, separated by pieces of flannel moistened with brine. The first modification of this voltaic pile was due to Cruickshank, who suggested placing the plates vertically in a vessel filled with a solution of brine. Dr. Wollaston in 1815 increased the surface of the copper plates in order to minimise polarisation, and Offerhaus and Hare devised a spirally constructed cell in 1821. In this last arrangement the plates of copper and zinc, separated by strips of india-rubber or vulcanite, are bent round into a spiral roll, and the active surface, for the amount of space taken up, is thereby much increased. This cell is capable of giving a large current, but the surface of the zinc soon becomes covered with zinc sulphate, which greatly increases the resistance and produces internal short circuits; these defects soon caused this cell to be abandoned. Single-fluid cells, in which acidulated water is the only fluid, have long since fallen into disuse, being replaced by other more convenient systems, which we shall now describe in detail.

Copper sulphate cells.—The principle of this cell was first suggested by the French physicist Becquerel in 1829, but Daniell was the first to practically realise it. The negative pole of the battery is a rod of zinc, lightly amalgamated and immersed in a 10% solution of sulphuric acid contained in a porous pot. The positive pole is a copper cylinder surrounding the porous pot, the whole being placed in a glass vessel containing strong copper sulphate solution. The copper sulphate absorbs the hydrogen deposited on the copper plate, the result being that pure copper and not hydrogen is deposited on it. The electromotive force of this cell is approximately one volt, and its internal resistance is anything from one-fifth to several ohms, and it is consequently incapable of producing large currents. As long as the copper sulphate solution is concentrated there is no polarisation, and its E.M.F. is therefore very constant. Owing to this constancy, this type of cell is much used where large currents are not required, as for instance in telegraphy or for electric clocks.

This cell has been improved and modified from time to time by various inventors. In the Calland cell the porous pot is dispensed with, and crystals of copper sulphate are placed at the bottom of the glass vessel, and cover a copper plate to which is soldered a wire covered with gutta-percha. A circular disc of zinc is suspended from the top of the vessel and is immersed in the pure water with which the vessel is filled; the manipulation of the battery is thus considerably simplified. In the above cell the two liquids are only separated by their difference of density. Minotto improved on this by introducing a layer of coarse sand above the copper sulphate, which separates them more effectively. The Vérité cell (Fig. 24) is provided with a glass sphere placed at the top of the vessel, in which are placed crystals of the sulphate which serve as a store to keep the sulphate solution thoroughly saturated.

M. Jeanty has devised a Daniell cell, called by him the Fulgur, consisting of a square wooden box, very carefully made, and lined on the inside with copper. This vessel contains the depolarising solution, and the copper lining is the positive electrode. Placed in it is a second receptacle containing crystals of the depolarising copper sulphate, and also a porous pot containing the zinc and the active liquid. Each cell has an orifice near the top, opening into a gutter, which prevents the liquid from rising higher than a certain level. The receptacle containing the crystals of copper sulphate is pierced near its base with holes, in order to allow the saturated solution of copper sulphate to flow out into the cell, when the lighter liquid containing the products of the chemical action in the porous cell are drawn off at the above-mentioned orifice. The porous pot acts as an osmotic diaphragm, and allows the solution, charged with zinc sulphate, to escape into the outer vessel by the simple process of osmosis. This weak solution of zinc sulphate is lighter than the saturated solution of copper sulphate, and rises to the top, where it may be drawn off. The liquid in the outer vessel consists at any instant during the action of the cell of a series of superposed layers, whose density gradually decreases on ascending to the top. Those at the bottom consist of saturated copper solution, while near the surface the liquid is very nearly pure dilute zinc sulphate solution. The flow of the saturated solution of copper sulphate into the outer vessel is regulated in various ways, by different instructors. From the large sizes of this cell currents up to 10 ampères may be drawn at an effective pressure of one volt. The current is as near as possible constant until the plate of zinc constituting the negative pole is completely consumed. This plate is about ⅓ in. thick when new, and will last several hundred hours before it is necessary to replace it. It is sufficient to attend to the battery once a week, and to thoroughly clean it out once a month, in order to ensure regular and constant generation of the current.

Cells using nitric acid.—The first cell using nitric acid as a depolarising agent was invented by Grove in 1839. It consists of an outer cell of glazed porcelain or glass, containing an amalgamated zinc plate and dilute sulphuric

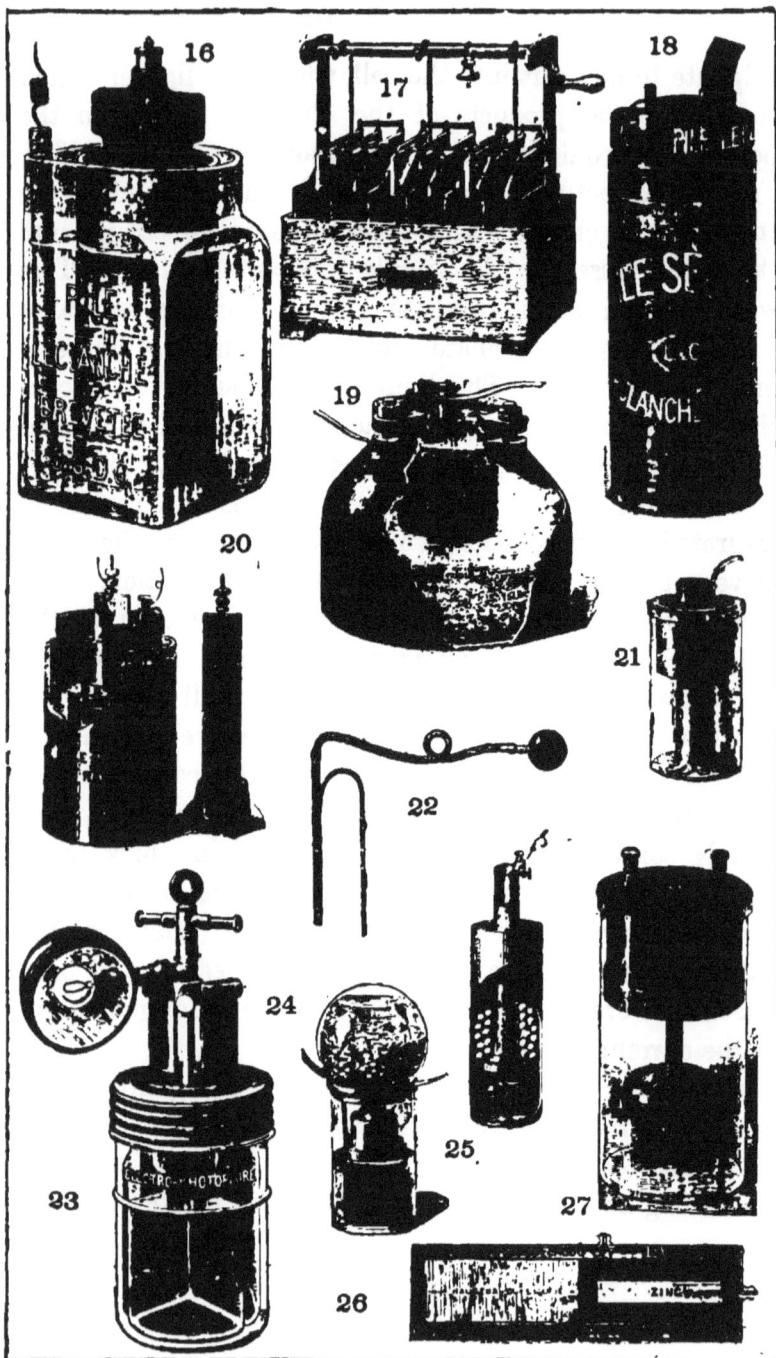

FIGS. 16—27.—Primary Cells.

acid. In the inner porous cell a piece of platinum immersed in concentrated nitric acid serves as the positive pole. The E.M.F. of this element is high, being about 1·9 volts, with a low internal resistance, which allows a large current to be obtained without there being much difference between the E.M.F. of the cell and the potential difference of its terminals.

A large number of different arrangements of this cell have been invented, notably the Bunsen element (Fig. 20), in which the expensive platinum plate is replaced by a block of gas carbon; also the Calland cell, in which the porous pot contains a solution of saltpetre, concentrated sulphuric acid, and nitric acid. Tommasi closes up the porous pot by means of a plug of paraffin wax, which prevents nitrous fumes from being given off into the air, but at the same time sensibly diminishing the E.M.F. D'Arsonval places the carbon outside and the zinc in the porous pot. The Ruhmkorff cell is rectangular in shape, the depolarising agent being a solution of potassium or sodium bichromate in a mixture of nitric and sulphuric acids. The shape of the cell and the composition of the depolarising agent is capable of almost infinite variety.

Bichromate cells.—The bichromate battery, which is due to the chemist Poggendorf, is constructed both as a single and two-fluid element. In both cases the positive pole is formed by one or more plates of gas carbon, either sawn in the crude state, or pounded and compressed into the right shape. The negative pole is a plate of amalgamated zinc. In the single-fluid type the liquid is a solution of sodium or potassium bichromate in water, to which is added sulphuric acid in varying proportion. The best-known type of single-fluid bichromate cell is the bottle cell of Grenet (Fig. 28), which is very useful when a current is wanted for a short time, as in the laboratory, or for medical electricity. Batteries of four to ten elements placed side by side (Fig. 17), with a small windlass or other apparatus for raising the zincs all together out of the

Fig. 28.—Bichromate Cell.

liquid when the battery is not in use, are very convenient where a higher E.M.F. is required. They are generally filled with a mixture of about 200 grammes of bichromate and 400 grammes of sulphuric acid per litre of water, and will generate about 20 to 24 watts per second per cell, for about $3\frac{1}{2}$ hours. There are a host of other cells of this type, which differ only in their shape and in the composition of the liquid mixture.

In the two-fluid cells the amalgamated zinc is kept separate in a porous cell containing dilute sulphuric acid, which is itself immersed in a glass or porcelain vessel containing an acid solution of bichromate of potash, which acts as a depolariser. Polarisation and the liberation of hydrogen are more thoroughly destroyed in this type than in most of the others, and the current it furnishes is consequently more constant, but, on the other hand, the internal resistance is considerably increased by the porous pot, which is a disadvantage.

Many attempts have been made to obtain a longer duration of the current, or to produce a greater output with the same constancy of E.M.F. Among cells invented with this object in view, we may mention the non-polarising cell of Cloris-Baudet, which has an internal porous pot containing sulphuric acid and bichromate crystals; the siphon cell by the same inventors, which is provided with india-rubber siphons for causing the dilute acid and the bichromate to circulate through all the elements of the battery, so as to completely exhaust the active materials; also the circulating batteries of Siemens, Hospitalier, and others, in which the cells are arranged one above the other in tiers, the liquid passing through each cell in succession, so as to make sure of its parting with all its oxygen at the electrodes; finally M. Radiguet's constant two-fluid cell, which is one of the most convenient and economic. In nearly all the bichromate cells it is necessary to lift up the zincs out of the acid when they are not in use, otherwise chemical actions go on, which would very soon exhaust the batteries. M. Radiguet has invented several ways of getting out of this difficulty, so as to always have the battery ready for instantaneous use. In one of them the zinc dips its end into a small basin of mercury made of paraffined wood; in another (Fig. 27) there is a small platform at the bottom of the porous pot, on which can be heaped granulated zinc or any old pieces which have been discarded from other cells.

M. Radiguet has also arranged a siphon (Fig. 22) for changing the liquid in any cells. The siphon draws up liquid into itself by squeezing an india-rubber ball, and discharges it on again compressing the ball. The cells can be thus easily re-charged with liquid without moving them from their position.

Potassium bichromate has been advantageously replaced by its active agent, chromic acid. The action of the cell is thereby considerably strengthened, and from ·25 to ·4 ampère per square centimetre of zinc surface can be obtained from cells using it as a depolarising agent.

A very light form of bichromate cell was devised and used by M. Renard for some experiments on navigable balloons at Meudon, consisting of a cylinder of platinised silver or a tube of carbon forming the positive pole, and an ordinary zinc rod forming the negative pole. The liquid used was a solution of chromic acid in hydrochloric acid, with sulphuric acid added when it was required to increase the output of current. About 15 litres or about 3 gallons of the mixture were required to produce one horse-power, and the total weight of the battery was 50 lbs. per horse-power.

This is about as light as it is possible to make primary

cells, but even with this exteme lightness, the expense of working and maintenance is so great, as to render them unfit for any but laboratory experiments, or for generating very small currents, such as are required for electric bells, telephones, etc. The cost of lighting a 16 candle-power incandescent lamp by primary cells would be about 6d. per hour, an expense which forcibly limits the applications of this type of generator of electricity.

Intermittent batteries.—To this class belong all cells which cannot be used for more than a few minutes together, after which they become polarised, and require to be left to themselves to recover. The best-known type is the battery invented by M. Leclanché (Fig. 16), which is universally used for electric bells, telephones, and for other domestic purposes, in which a battery is required for only a few moments at a time, and which shall always be ready for immediate use, and not waste away between whiles. The cells require very little attention, and the cost of their maintenance in working order is negligible. Since its invention, this form of primary cell has been considerably modified and improved, and several different types exist. Usually the positive pole is a mixture of coarsely-pounded gas carbon and manganese peroxide, enclosed in a porous pot, and the negative pole is a zinc rod. The porous pot is sometimes dispensed with, and the manganese applied in a conglomerate block to the face of the carbon. The liquid is always a solution of sal ammoniac in water. The electro-motive force of a new Leclanché is about 1·5 volts, and its resistance from ·2 to 2 ohms, according to its size. In consequence of the high internal resistance, only very small currents can be drawn from it, and even that for only a short time, so that it is limited to the uses above-mentioned. As the cell is in other respects very constant, and only requires re-charging about once every three or four months, it is very well adapted for domestic purposes.

Other cells have been devised on something the same principle as the Leclanché, using sal ammoniac or some other non-acid liquid, but capable of generating larger currents. The Warnon cell contains a cloth bag filled with coke and

FIG. 29.—Goodwin's Carbon (section).

peroxide of manganese; the Maiche element uses the oxygen of the air as a depolariser; Goodwin makes use of corrugated carbons so as to increase the surface of the positive pole (Fig. 29), in order to keep the resistance as low as possible. None of these types have been commercially successful, and the Leclanché cell, with a porous pot or conglomerate blocks, is practically the only one used at the present date.

Dry batteries (Fig. 18).—This type of cell, in which the depolariser is solid, and the liquid is soaked up in some absorbent material, has lately come into favour, especially in

Germany, but as a rule they have the disadvantage, that when once the materials are exhausted, they cannot be replaced, and the cell is of no further use. The best dry cells are constructed by Siemens.

Thermo-electric generators.—If two different metals be joined together by solder, and the junction of the two heated, it is noticed that an electro-motive force is generated at this point. Many inventors have attempted to make use of this phenomenon to produce electric currents by the direct application of heat. The discovery of thermo-electricity dates from 1822, and is due to Seebeck, and since then thermo-electric generators, or thermopiles as they are called, have been devised by various inventors. The best known are: the thermo-electric multiplier of Melloni, which is an instrument of extreme sensitiveness, and capable of measuring very small differences of temperature. Clamond has constructed powerful thermopiles of iron and galena; others have been invented by Noé, Jacques, etc.

The thermopile consists of a large number of small pairs of two different metals, usually antimony and bismuth joined together in a series. Alternate joints are heated while the remainder are kept cold. Thus all the antimony-bismuth joints starting from one end of the chain are heated, while the bismuth-antimony joints in the same direction are kept cool. As a generator of electricity, the efficiency of the thermopile is very low, probably not more than 1% of the total heat supplied to it being turned into electricity. It has been calculated that even in the best types, such as the Chaudron generator, over 1000 cubic feet of coal-gas would have to be burnt to produce one horse-power hour, whereas a good gas-engine only burns about 20 cubic feet in order to do the same amount of work. For this reason, this class of apparatus has been confined to laboratory instruments, and will probably never take a place in the commercial generation of electricity.

CHAPTER VI

ACCUMULATORS

An accumulator is an apparatus in which electrical energy may be stored up as chemical work, to be given out again when required. The action of an accumulator may be compared to a reservoir into which water is pumped at a more or less rapid rate, and from which it may afterwards be drawn, at a constant rate, which may even exceed that at which it was filled. In the accumulator the current of electricity which charges and discharges it represents the stream of water; but we must not push the analogy too far, for it is not electricity which is stored up in the accumulator; the electrical energy is converted into chemical work in the cell, producing various chemical actions. Thus when an electric current is passed through lead plates immersed in dilute acid, the metal is oxidised by the phenomenon of electrolysis, and a small portion of the positive lead plate is converted into peroxide of lead, whilst the negative plate becomes spongy. If the current is stopped, and the plates joined by a conducting wire, a current flows through it in the opposite sense,

61

which lasts until the plates have returned to their original condition.   While the current flows, the chemical energy which had been produced by the original *charging* current is converted back into electric energy.

Ritter was the first to construct a secondary cell, but he only succeeded in getting out of the battery a very small portion of the energy which he put in.

Gaston Planté constructed the first really practical accumulator, consisting of two sheets of lead rolled up without actual contact, dipping into dilute sulphuric acid. He *formed* the plates by sending a current through them, first in one direction and then in the other, a great number of times, so as to make them capable of preserving and discharging the energy stored up in them.   M. Faure perfected this first accumulator by reducing the time required to form the plates, which at first took several months of the charging and discharging process.   M. Faure covered the plates with a coating of red-lead, which rapidly becomes peroxidised at the anode, and reduced to spongy lead at the cathode, when a current is passed through the cell to charge it.   This was the first step towards *forming* the plates artificially, and established an entirely different type from those formed naturally.   Both methods of forming the plates are in use at the present day, some makers preferring the artificial, others the natural method of formation; others again combine the two.   We will now describe a few of the best-known types.

Tudor accumulator.—This accumulator is founded on Planté's invention, but has been modified, in order to make the perfect cell of to-day.   It is found that the accumulation of energy is proportional to the depth of the layer on the plates, which is oxidised. It must be remembered that the repeated oxidation of the metal and its return to its first state, which constitutes the *formation* of the plates, are operations which proceed very slowly, and the length of time taken by the process is quite incompatible with the rules of rapidity and economy which regulate, or should regulate, commercial industry. The Tudor system of accumulators was designed to avoid this inconvenience as much as possible, and is largely used on the Continent and at home for central stations. The principal characteristics of the Tudor accumulators are their mechanical strength, their very great active surface, and the small quantity of oxide necessary for their formation. The skeleton of the plate is cast with a surface, scored with horizontal grooves. The plate is first formed by the Planté process, and then the grooves are filled with oxide and the plate formed afresh.

It is found that in all accumulators, the oxides of lead, laid on by the Faure process, tend to scale and drop off to the bottom of the accumulator after a time. In the Tudor accumulator the pasted oxide drops off also, but there is left the underneath plate, which was originally formed by the Planté process, and which, in the course of the usage which it has undergone, has much increased in capacity.

The Tudor secondary cell was designed principally for central stations, and is constructed so as to be capable of a much higher rate of discharge than the normal, without risk of injuring the plates. The electrodes are kept as far as possible apart, and the bottoms of the vessels have been arranged so that any scales or dust of oxide which may drop off the plates shall not, under any circumstances, be productive of short circuits.

Accumulators are very seldom moved after being once placed in position, and so the connection of one cell to the next is often made by melting together the *lugs* of the two cells : the piece of lead to which all the plates in a cell of the same sign are joined is spoken of as a *lug*. This melting together of the lugs ensures good electrical connection, which is of the greatest importance in practical working.

As a rule, each cell is required to furnish a current which would be too great for a single positive or negative plate without making it inconveniently large, so that a number of smaller plates of each kind are put into a cell, and each set connected together to a lug. There is as a rule one more negative than positive plate, the positive plates being sandwiched in between the negative plates. The cells contain as a rule 5, 11, 19, and 33 plates depending upon their *capacity*, which is measured in ampère hours. The Tudor Company guarantees its cells for ten years, with a depreciation of $3\frac{1}{2}\%$ of the prime cost.

Faure-King accumulator.—This accumulator (Fig. 30), invented by Mr. C. A. Faure and Mr. F. King, and constructed by the Electrical Power Storage Company, represents the most recent developments of the storage cell. The plates belong to the type of pasted or artificially formed plates. The frame-work of each plate consists of an alloy which plays the combined part of conductor and support for the active material, and is only slightly expansible. Each plate, whether positive or negative, is inserted into an envelope, composed of a special form of celluloid, which resists destructive action on the part of the oxides, the gases, and the electrolyte. A further support

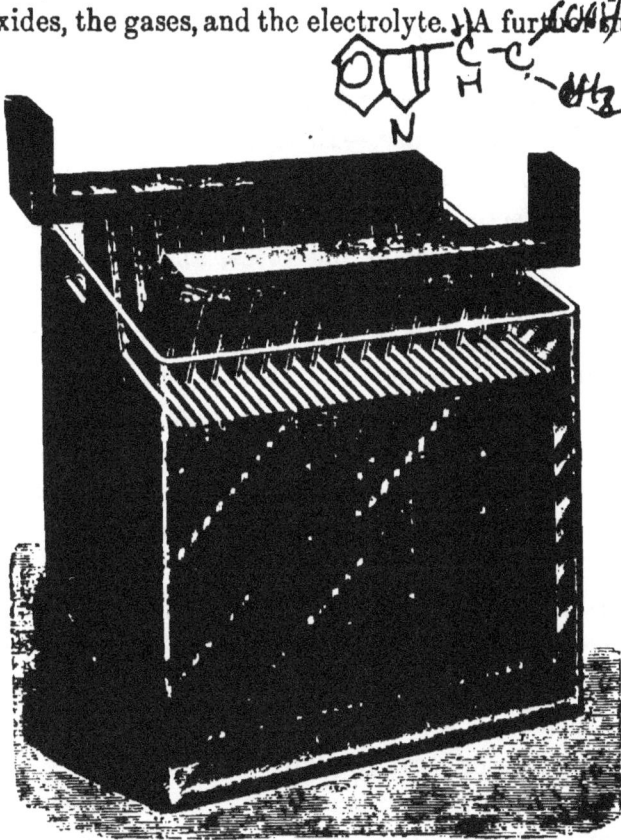

FIG. 30.—Faure-King Cell.

is provided in each plate by a number of studs, the stems of which pass through the body of the plate, and thus form double-headed bolts holding the surfaces of the envelopes and the material contained in them firmly together.

F

These cells are made in various types according to the uses to which they are to be put. In the traction type the approximate complete weight per kilowatt-hour spread over five hours' discharge is about 140 lbs., at a discharge rate of 15·5 ampères per square foot of plate.

Dujardin accumulator.—Except for the No. 1 type of 15 ampère hours' capacity, in which glass vessels are used to hold the electrolyte, all these accumulators are contained in metallic boxes, cast in one piece, and provided with a movable lid. The plates are of the Planté type, and are built up of a number of lead strips. The positive plates are made slightly larger and thicker than the negative, and the normal discharge rate is about 10 ampères per square foot of positive plate. The sides of the vessels have a transparent pane let into them, in order to enable one to see at a glance the height of the liquid within.

Epstein accumulators. — The plates are also of the Planté type, and are prepared by boiling in a 1% solution of nitric acid and water, which renders the lead porous. In addition to this the plates are finely corrugated so as to greatly increase the surface. The normal rate of discharge is 30 ampères per positive plate.

Crompton-Howell accumulator.—The surface of the plates of this accumulator when unformed has a crystalline appearance in consequence of the peculiar process of making them. The lead alloy is melted, and then allowed to crystallise by slow cooling, and the plates, being sawn out of the solid crystalline mass, present a peculiar appearance. This process renders them porous to start with, and they are then further formed by the ordinary Planté process. Considerable space is allowed between the plates for free circulation of the electrolyte.

Faure-Sellon-Volckmar.—The illustration on this page

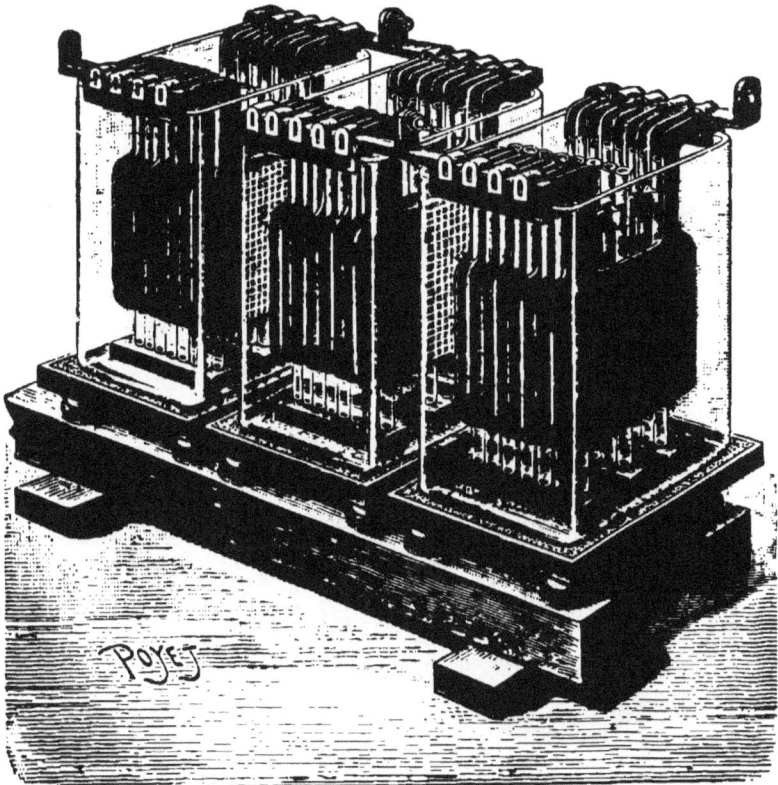

FIG. 31.—Faure-Sellon-Volckmar Cells constructed by Valls et Cie.

gives a very good idea of these cells, which have been specially designed to obviate most of the difficulties found in the employment of pasted cells. The plates are suspended at a considerable distance from one another and from the bottom of the vessels, so that there is very little fear of scale and particles of oxide making short circuits between them. Glass tubes are inserted between the plates to make sure that they are held at the proper distance apart.

Fɪɢ. 32.—Cells of the Société du Travail Electrique des Métaux.

Accumulators of the Société du Travail Electrique des Métaux.—The plates of these accumulators appear divided into small squares (Fig. 32), in consequence of the peculiar method of their construction. A chloride paste is first cut up into small blocks, which are placed in position, and molten lead is run round them so as to form plates. These plates are then heated by the ordinary Planté process, the chloride being first reduced to spongy lead for the negative plates, and then oxidised into peroxide for the positive plates. These accumulators have been used for traction in the Tramways-Nord of Paris; they have large capacity, and their efficiency is very high.

We may also mention, of the systems now in use, the Blot, Vernon, I. E. S., and other accumulators.

Management of accumulators, charging, discharging, etc. —The liquid for filling the cells must be distilled water, to which sulphuric acid is added until the specific gravity of the mixture is 1·190 when cold. The sulphuric acid must be free from impurities, such as arsenic, nitric, or hydrochloric acid. When the cells are fully charged the specific gravity should be from 1·20 to 1·22.

The acid solution should be put into the cells to a height of not less than half-an-inch above the tops of the plates, and the level should be kept constant either by adding pure water or weak acid until the specific gravity is that stated above.

Charging.—A constant current is not absolutely necessary for charging, but it should never exceed the value for which the cell is constructed by the makers. The charge is not complete until violent ebullition of the gases evolved has proceeded for some time. When this is the case, the potential difference between the terminals of charging cells ceases to rise, and each cell has a terminal voltage of 2·5 to 2·75 volts, according to the charging current.

Surcharge.—The battery should always be kept as fully charged as possible, and should as well be charged to a sufficient extent to cause the liquid to become milky at least once a week. If the battery has been out of regular use for some time it must be *surcharged* for about two or three hours more after it is apparently fully charged. It must be remembered that if the plates are in good working order, surcharging too frequently may become an abuse, and be actually harmful to the cells. It is best, if practicable, when the cells are to be out of work for some time, to leave the battery fully charged, and then to give it a short charging, say once a fortnight, till the acid turns milky, by which treatment the cells will be kept in order for any length of time.

Discharge.—The normal maximum rate of discharge should not be exceeded, and should never be continued after the specific gravity of the liquid has fallen below 1·17, or the terminal voltage of a cell below 1·8 volts, at the full rate of discharge. Under no circumstances whatever must these limits be exceeded.

Re-charging.—After each discharge, either partial or complete, the cells should be immediately fully re-charged, with the least possible delay. Not more than twenty-four hours at the outside should elapse between the end of the discharge and the beginning of the re-charging.

Disuse.—As stated above, if the battery is to be left out of work for any considerable time, the battery must be fully charged, and kept in this condition throughout the period of disuse. When starting them again, the cells should be first surcharged for two or three hours.

Towards the end of a charge, at the moment when ebullition begins, the attendant in charge of the cells must examine each cell carefully to see whether gas is being evolved equally from all the cells. If, during this necessary but simple inspection, any cell should be noticed which has not reached the same stage as the others, or has not commenced ebullition at all, it is probable that such a cell has become internally short-circuited by pieces of paste or scale which have fallen down between the plates. These should be removed with a thin lath of clean wood, and the cell should be cut out of circuit during the discharge, by disconnecting one terminal of the cell, and connecting the two adjoining cells by means of a piece of cable large enough to carry the discharging current. The cell should be restored to its position when charging, and it will generally be found that one or two charges restore the cell to its proper condition.

These are the principal rules which *must* be rigidly observed in the maintenance of all accumulators. If carefully and methodically looked after, using only the smallest discharging current which is absolutely necessary, the accumulator will be found to be an efficient and economic piece of apparatus, whose utility in all well-equipped and really complete installations is beyond dispute.

As a regulator of pressure and reservoir of electricity, the accumulator is capable of rendering great service, and its value is proved by the ever-increasing use which is made of it in all electrical generating plants. Even the most modest of generating stations realise the necessity of including a battery of accumulators in their plant, as they not only increase the output of the station, but also its efficiency and economy, as the following reasoning shows.

In everyday life light is required only for about six hours in winter and about four hours in summer, consequently in a generating station, during a greater part of the twenty-four hours, only a very few of the total number of lamps which the station supplies are alight, which it would be very wasteful to keep alight by means of a powerful engine and dynamo. This is where the advantage of accumulators comes in : they may be either charged during the day-time, at the same time supplying the few lamps alight, and then help the dynamo during the evening to cope with the heavy load which is thrown on the station at night-fall, or else they may be charged during the time of maximum load, by a dynamo large enough to supply current to the lighting circuit at the same time as charging the accumulators. By this last method the dynamo and engine are working efficiently under a heavy load for a few hours, while the rest of the time the engine is shut down, what little load there is being thrown on the accumulators.

This idea of using the accumulators for increasing the economy of electric installation is old, and numerous more or less successful systems have been devised with a view to this object. In small out-of-the-way stations, secondary cells are absolutely necessary, especially if, as is sometimes the case, the motive power is irregular or intermittent ; for example, water or wind power. In these cases the accumulator stores up the energy when available, from which it may be drawn when required at a totally different and constant pressure.

The only disadvantage to accumulators is their price, which is high, and forms a considerable portion of the capital outlay in stations reckoning the weight of their accumulator plates by tens of thousands of pounds.

In a large installation, the plant consists perhaps of a couple of engines and dynamos, and a battery of accumulators. The usual working conditions are as follows:— During the day the load on the station consists of a few lamps and perhaps motors making in all say 50 horsepower. At night this load becomes very much larger, and may rise to ten times the normal day load, meaning in this case five or six hundred horse-power. The whole demand might be met by engines and dynamos of 600 horse-power but this would leave them working inefficiently during the greater part of the twenty-four hours, so that it is better to have a somewhat smaller engine, charging accumulators during the day and supplemented by them during the evening, or a larger engine taking the whole of the evening load and charging the accumulators as well. During the day the engine would be at rest and all the current would come from the accumulators. In most cases, therefore, even when a constant and regular supply of waterpower is available, accumulators form a natural part of a well-equipped station. They are to the electrical engineer what the gasometer is to the gas engineer, a reservoir which can always be drawn upon. This explains, from a commercial point of view, the great success which has attended them, since inventors applied themselves to making them really economic and practicable.

CHAPTER VII

As a whole volume of this series will be devoted to the study of dynamos, we will not go deeply into the subject, but confine ourselves to giving a brief account of the electro-magnetic principles on which they are designed and constructed. We will also describe the general features of the best-known types.

Historical.—Faraday noticed that every time a metallic wire was moved in a magnetic field, produced either by a current in another wire or otherwise, a current was produced in it whose effect was to resist the motion. This fact led him to enunciate the laws given in the third chapter of this volume. Ampère also showed, that the magnetic field produced by a permanent magnet could be artificially reproduced by employing a current flowing through a spiral of wire, and that the strength of the current induced in a wire moved in this field depended on the length of the wire, the rapidity with which it moved, and the strength and direction of the magnetic field. We therefore see by the above phenomena that the mechanical work expended

74

in moving a wire can be converted into electrical energy in it. These discoveries suggested the construction of magneto-electric machines, in which a wire should be continuously moved in a magnetic field by some mechanical contrivance so as to produce an electric current.

The principle of a mechanical generator of electricity is therefore simple. All that is necessary, is to cause a closed coil of insulated wire to rotate in front of a magnet. If this is done, a current is produced in the coil which flows first in one direction round it and then in the opposite direction. If such a dynamo were required to generate a current, always in the same direction, it must be provided with a ring split in two halves, rotating with the coil, and each half connected to one end of the coil. Against this ring rub two plates of metal or brushes one on each side, to which are attached two wires leading to the circuit in which the current is required.

The action of this *commutator* is as follows:—At the instant when the current in the coil is just reversing, each brush is passing from a position touching one half of the ring, to a position touching the other half, which its fellow has just left. The effect of this is that the external wires, mentioned above, are now connected to ends of the rotating coil opposite to those to which they were originally connected, so that the reversed current in the coil still flows in the same direction in the external part of the circuit as it did before. This direct current is obviously not continuous, but varies from a maximum to zero. If, however, this variation is rapid enough, the current becomes practically continuous.

The first machine built on this principle was constructed in 1832 by Pixii; it consisted of a movable magnet, spun rapidly in front of two fixed bobbins of insulated wire Saxton and Clarke reversed the arrangement and spun the bobbins in front of the magnet. In 1849, Professor Nollet constructed a powerful magneto-electric machine with large horse-shoe magnets, between the poles of which moved coils of copper wire furnished with iron cores Masson and Van Malderen perfected this machine, and i was brought into practical use for lighting lighthouses by electricity.

The rotating coil is called the *armature*, and in 1856 Siemens devised a more perfect arrangement, in which the coils of insulated copper wire were wound lengthways along a spindle-shaped core. This armature, whose cross section is something like an ⬚⬚, and forms a powerful magnetic core, was rotated between the poles of a series of adjacent field magnets. The next improvement, due to Wilde, was the employment of the current generated by Siemens' machine to magnetise a powerful electro-magnet which in its turn formed the poles of a second larger dynamo machine. In 1866 Siemens and Wheatstone independently suggested leaving out Wilde's small magneto machine, because the iron which constituted the poles of the second machine contained enough residuary magnetism to generate a small current in the armature which current, being directed through the coils forming the electro-magnet, would increase its magnetism, thereby causing an increase of current in the armature, and so on This process was found to continue until the iron of the core of the electro-magnet, which is called the *field magnet*, becomes magnetically saturated, that is to say, incapable of further magnetisation.

This is practically the principle of the direct current dynamo as it stands to-day. The next greatest advance in the construction of dynamos, was the invention of the ring armature by Gramme, who patented it in 1870.

This armature consists of an iron ring-shaped core made of a bundle of annealed iron wires. Round the circumference of this coil are wound transversely the armature coils, each made of the same number of copper wires and insulated from one another. The end of one coil and the beginning of the next are connected to one of a number of metallic strips, laid round the circumference of a collar of insulating material, which is fixed to the shaft of the dynamo, the number of strips being the same as the number of coils, the whole forming the *commutator*. Such a ring armature may consist of 20, 40, 60, or more coils according to the size of the machine, and turns between the poles of powerful field magnets, which are made to fit as closely to the rotating armature as possible by means of pole pieces (Fig. 33). The current is collected from the commutator by means of *brushes* of copper gauze which rub on it, separated from one another by 180°, and connected to the terminals of the machine. These brushes are gripped by *brush-holders*, which in their turn are held in position by a light frame-work called a *rocker*, by means of which the brushes can be set in any required position round the axis of the commutator.

The generation of the current in a dynamo is governed by the laws of induction, and consequently the electro-

www.ingramcontent.com/pod-product-compliance
Lightning Source LLC
Chambersburg PA
CBHW021954190326
41519CB00009B/1260